I0483062

App Sense
Understanding and Using Apps for Kindle
Fire, Apple, and Android Devices

This book is written for my mom, Rita, and for anyone else who needs a little extra help understanding apps and how they work. Mom, thanks for giving me the idea for the book and the great title!

Other Books Written By Shannon VanBergen
Border Hoarder
The Accomplishment Journal
Feeding Time – A Children's Book about Animals

First Printing, 2014

ISBN-13:
978-1497587779

ISBN-10:
1497587778

www.WhatAboutApps.com

Disclaimer – The app world is constantly changing. Some of the apps that I use for examples may change or eventually not be available.

Table of Contents

Chapter 1 - Introduction

In 2009 Apple came out with the catchy slogan "There's an app for that". At the end of 2009 Apple had approximately 100,000 apps in their store. Now, just four years later, there are over 1 million. That's a lot of apps! You can find apps devoted to chicken recipes or apps that help you trouble shoot problems with your vehicle. There are even apps that will help you block people's phone numbers on your phone so you don't text them or call them when you've had a little too much to drink (yes, they actually exist).

For those of us who have been buying and using apps all this time, apps have become a daily part of our lives. We check our bank accounts, check the weather, check our flights, and play games when we're standing in check out lines. We love our phones and tablets so much we know our parents or even grandparents will love them just as much, if not more. But for those just starting out, apps can be daunting, especially for those not used to all of this technology.

This book will help you understand how to search for and purchase apps, how to delete apps on your device, how to set up parental controls, how to protect your privacy and much more.

I've written this book in a question and answer format. Instead of having to read the entire book (but you should because it's full of great information) you can simply find the question you need answered in the table of contents.

Chapter 2 – What is an App?

The word app is short for "application". Unlike computer programs, which allow you to do several things, apps usually only preform one (or very few) tasks. One app will let you check the stock market, another will let you check the weather, and another will let you play a game.

Chapter 3 – Do I Need an Internet Connection to Use my App?

Even though you will need an online connection to download an app, you usually don't need it to use the app. For example, you might download an app that has recipes on how to make your own cleaning supplies. Once you've downloaded your app, you don't need an Internet connection to look at the recipes. They are saved in the app on your phone.

However, there are some instances when you will need an Internet connection. See the examples below.

1) **You download a game app (like Bejeweled).** You don't need the Internet to play the game but you will need it if you want to post your score on Game Center (for Apple games), Google Game Play (for Android) or Game Circle (for Kindle). If you don't want your high scores posted, or to see other people's high scores, you won't need an Internet connection. Most games will continue to keep track of your high score in the app without an Internet connection.

   If seeing other people's scores, seeing your score posted, or seeing how you stack up against other players adds to the fun of the game, then by all means, post your score. On some games I post my score and on others I don't.

2) **When an app needs updating**. Sometimes an app will need to be updated and to do that you will need an Internet connection. App developers sometimes

tweak the game to make it better or they will fix certain "bugs" to make it work better. You will get a notification that a certain app you have downloaded has an update waiting. You can either chose to ignore it, or to update it. The notification will let you know what it's updating. For example it might say that it's fixing a bug that caused it to crash or to be slow. Or it might say that it's adding new features. Most of the time the new updates are good, but sometimes they are not. It's up to you if you want to update the app or not.

3) **The app might contain links to a website**. Apps can only hold so much information. Some apps, like a sports app, are mainly used as a way for sports fans to see the schedule of their favorite team. If the developer would like to share more information with you, like the latest news of this sports team, or give you a chance to post your opinions on Facebook on how they are playing, they will put in a link to take you to a certain website from the app. You will need an Internet connection to use this.

Chapter 4 – What Should I know Before I Install an App?

With all the apps out there, many of them being free, it's tempting to start downloading right away. Here are some things you need to keep in mind before you start adding to your app collection.

1) Before you can purchase an app you will need to open an account with Amazon, iTunes or Google Play, depending on your device. When you go to their website they will walk you through, step by step, how to sign up for an account.

> Here are the websites to open your account:
> For Kindle – http://www.amazon.com/
> For Apple -
> http://www.apple.com/itunes/download/
> For Google Play - https://play.google.com/store

Note – When you sign up for an account it will ask you for your email address and password. This doesn't mean that they want the password to your email address. They will use your email address as your log in name and they are asking you to set up a new password for this account.

2) Apps take up room on your smartphone or tablet.
Your phone or tablet can only hold so many apps. Each app will tell you in the "Descriptions" section in the app store how large the app is. Phones and tablets make it easy to tell how much storage you have left on your device. If you're getting low on storage and want to download more apps, you may have to delete apps already on your device. See

the chapter on deleting apps in this book to learn how to do this

3) When you download an app from any app store it requires two things from you – trust and common sense.

Before you buy any app make sure to do the following:

Read over the comments and reviews other customers have left.
This is where you can see if other people are having problems with the game itself or if the game has caused problems with the person's phone or tablet. You will want to read several of the comments, not just the first few.

What is the star rating?
If it has a low rating it could mean the developer threw a game together with the purpose of getting information from users that they could use maliciously. However it doesn't always mean this. If there are only a few bad reviews it will drive the rating down. A competing developer who wants to sabotage a game it sees as a threat might have left the bad reviews there. Sadly, this happens. Also new games won't have any stars yet or very few comments.

Check other games the developer has made.
This information is usually available in the "Description" section of the app on the app store. Check out what other games they've made and what kind of reviews they've received. What you're looking for is anything that might hint to this developer using your information in a way they

shouldn't or making apps that contain malicious content or viruses. In other words, look for red flags.

Check the permissions the game asks for.
Some apps require certain information from you. For example if you're downloading a photo-editing app it will ask for permission to see your pictures. This doesn't mean that your pictures are sent to the developer. It just means that the app will need to open your photos to retrieve whatever picture you want to work on.

However, some permissions aren't in line with what the app does. For example there is a calculator app on the Android market that wants to know your exact location. It shouldn't need that to add up your bills or your groceries. It's probably there so the ads on the app can target your local area.

If you come across an app that is asking permission to access information from you that you don't think it needs, then don't download it. However, keep in mind that many of the permissions sound a lot scarier than they really are.

For a list of commonly used permissions and what they mean see the chapter in this books that explains them.

Other things you should know:

* If you purchase an app by accident you can remove it from your device. See the chapter in this book that tells you how to do this.

* If you don't like an app that you downloaded, get rid of it and get a new one. There are literally over a million apps in the Apple store alone. If you download an app, and you don't like it, get it off your phone. Don't feel guilty if you spent money on it. Most likely you only spent a few dollars anyway. There is limited space on your device and it should be used for apps that you like and will use.

* If you delete an app off of your phone you still have access to it. You can reinstall it by logging onto your account online and looking through your purchasing history.

Chapter 5 - What are Permissions and why are There so Many?

Sometimes when you download an app it will ask for permission to access some of your information or to access parts of your device. If you deny them permission, you can't use the app. Some of the permissions can seem a little scary but are actually harmless. For example if a photo editing app asks to have access to your photos you might be concerned about giving them permission to use them. In reality they don't actually see your photos or store them to use later. The app just needs permission to access your photos so you can edit them in the app.

There are some permissions that aren't necessary and you do need to be vigilant of those. For example there is a calculator app in the Android app store that wants to know your exact location. That's unnecessary. The reason they probably want to know is so they can target ads to your location. If you think apps are asking for permission to see or use things you don't think are necessary, then look for another one that doesn't ask for so many permissions. Just keep in mind that certain apps will have to have access to some of your information so the app can run.

Below are a list of common permissions and why an app might need them.

Make Phone Calls – This allows the app to call any phone number without going through your regular dialing screen and without you needing to confirm the call. If the app is not a trusted app, don't agree to this. This can give the app the ability to call costly 1-900 numbers without your permission and leave you with the bill. The only kind of app that would need permission to do this would be apps like Skype, Google Voice, Google Talk and Google Maps. If the app you're installing doesn't have anything to do with calling people, then don't agree to this permission.

Send SMS or MMS – This one is a lot like the phone calls permission where you can be charged for sending texts (SMS stands for "Short Message Service which is a text and MMS stands for Multi Media Service which is sending pictures and videos) if it's from a malicious app. Make sure you read the reviews of apps that ask for this. You will be able to tell from the reviews if people are happy with the app or if they found out they were scammed. Legitimate apps that would require his would be texting apps or messaging apps.

Sometimes apps want this information so they can use it to send you a message if you make an In App Purchase. For example Angry Birds recently asked for permission to access your text messages so you could be sent a text showing that you made an In App Purchase. It used to be that they would bill your credit card for purchases but the developer of the game made it so you can make purchases and it will be added to your cell phone bill. Your phone provider would then send you a message letting you know you were billed.

Note - Even though angry birds asked all of their customers for this permission, the new billing change is only available in one country at this time. All other customers still charge their credit card. By asking for it all at once they are preparing for future billing changes.

If you're unsure why an app would want access to read or send you texts always find out why they need it. If it doesn't say on the app you can always Google it and many times you will find your answer there. If you don't feel comfortable with it or don't think it's legitimate don't download the app.

Modify or Delete SD card content – This one sounds really scary and it can be if you're downloading a malicious app since you're giving it permission to add, edit, read or delete anything stored on your SD card. As long as you're purchasing a legitimate app (one that has many good reviews, a good star rating, and/or comes from a reputable source) you will be fine. Many apps including camera apps, audio/video apps or document apps will need access to your SD card so they can store data on it.

An example of an app needing to change your SD card content would be a photo app. You download the app and make some changes to your pictures. After you make your changes you will want to save it to your SD card. The app will then save it to your card. Later you may want to change the photo editing on certain pictures (maybe turn them into black and white photos), the app will then need to edit the image on your SD card. Maybe you don't like the change you made so you will need to delete the picture through the app. By doing the above, the app will have added, edited and deleted something on your SD card. The developer of the photo-editing app never had access to your SD card. You were the only one who made the changes and authorized the changes.

Manage Accounts – This allows an app to find out what accounts you have and connect with them. And example of this is the Facebook App. It allows you to sign into your account through the app. If the app isn't account based (like Facebook, Twitter, etc.) then don't allow this permission. You don't want an app (like a game app) to be able to access your other accounts and delete them.

Read phone state and identify – This permission is used by many apps. It's usually used for game, music, or video apps or other apps that have noise. The app will need to know when a phone call is coming in so it can switch from the game audio (or music or video) to the phone call so you can hear the call ring.

Read Contacts – This permission is usually requested by social media apps like Facebook, Pinterest and Twitter, or apps that help you manage your contact list. If you're using the Facebook app you can ask it to search through your contact list on your phone or tablet to find your friends that are also using Facebook. If the app is not a social one (for a social media site or one that lets you send texts) or doesn't manage your contact list, don't let the app have access to your contact list.

Read/Write Calendar Data – This permission will allow the app to read and write in your calendar. Task manager apps and to-do list apps are an example of apps that will need this permission. Another example would be a birthday list reminder app. If an app that doesn't have anything to do with your calendar is asking for this permission, don't download it.

Read sensitive Log Data – This permission allows the app to read the log data of your other applications and this is none of their business. The only kinds of apps that should have this kind of access are apps that will debug your phone. If the app is not about debugging your phone, don't allow the app access to do this.

Read Browser History and Bookmarks – You don't want malicious apps being able to track your movement on your phone by checking out all the things you've looked up or bookmarked. Some legitimate apps might need this information (like apps that sync or back up your phone or tablet and some social apps) but most don't need it.

Coarse (network-based) location and Fine (GPS) location – These apps track your location. Some apps need this information, like Google maps. Other apps (like game apps) want to use it so they can send you location-based ads. Most of the time this type of permission is harmless and not used maliciously. However if you don't want apps to be able to track you, don't download it.

Full Internet Access – This is definitely one of those times when you need to check out the reviews before you download the app. Many legitimate apps need access to the Internet, but so do malicious ones. Check the reviews to make sure people aren't complaining about anything shady going on.

Prevent phone from Sleeping – This is actually a pretty helpful permission. Usually if you don't touch the screen on your phone or tablet for a few minutes it will shut down (or go to sleep) to save the battery. However this can be very irritating if you've downloaded the Kindle app and you're trying to read on your device. This permission allows the app to stay open until you close it out. This is also used for many game apps. This permission is not usually used maliciously.

Take Photos or Video – This permission is also usually harmless. Many photo-editing, camera or photo booth apps will require this permission. We have an app called 'Stache It that uses this permission. Within the app you are able to take a picture with your device and then add mustaches or lips or other crazy things to your photo. My daughter has a video-editing app where she can take a video and add in cool special effects like things blowing up or boulders falling from the sky.

Create Bluetooth Connection – This is another fairly harmless permission. It allows your phone to communicate wirelessly. I recently bought a Fitbit, which is like a pedometer that you can carry in your pocket or on a bracelet. Because it's small it doesn't show you how many miles you've walked. You have to download an app and, thanks the Bluetooth connection, it will sync with your Fit Bit and show on your app how many steps/miles you've walked so far.   Once in the app you can also log your other activities, and food and water intake. With out this permission I wouldn't be able to sync my Fitbit or use any of the features like setting an alarm or monitoring my sleep.

View network state (view Wi-Fi state) – This is a pretty harmless permission. It only tells the app if you're connected to the Internet via Wi-Fi or by 3G. Apps that need to download files on your device want to make sure you're in Wi-Fi mode before they do it. This permission can't change your Wi-Fi, it only checks to see if you're connected via Wi-Fi.

Push Notifications – When you first download an app it will ask if you'd like to receive push notifications from the app. This is not a text. It's just the app sending you a notification that something is happening in your game (if you're playing a farming game it can let you know when your crops are ready to harvest), that you're close to reaching your exercise goal (like my Fitbit does) or it can be used by the developer to send you messages about updates or other offers. Sometimes I accept them, sometimes I don't. This is the only permission you can turn down and still be able to access the app.

As long as you're reading the reviews and checking to make sure the app is legitimate and people aren't complaining about scams or spamming, you should be ok. If you're unsure why an app needs certain permissions, Google it. If you don't think an app should have that kind of access, don't download it.

There are many legitimate apps out there that won't do anything bad with your personal information and content on your device. Just use some common sense and you will be fine.

Chapter 6- How Do I Search For And Buy Apps on Apple Devices?

Searching for and buying apps in the Apple App Store is easy.

The following instructions may be a little different on each Apple device. For example, when I pull up the app store the "Categories" section is at the top on both my iPad and iPod but it simply says "Categories" on my iPad and on the iPod it says "All Categories". It's the same thing. Keep this in mind while I explain things. It may not look exactly the same on your device, but it should look very close.

Start by clicking on the Apple App Store icon. It's blue with a large white "A" in the middle. Once there you have a few choices in how you want to find an app:

1) Search Categories – If you're not sure which app you want this would be a great place to start. When you click on Categories you will get a drop down menu of all the different categories to choose from. You will see "Books", "Games", "Kids", "Health", and several more. You can click on whichever category suits you.

2) Best New Apps and Best New Games – This section is in the middle of your screen. Apple has chosen some of their more popular new apps to display here. You can scroll through from side to side to check out what's popular. Some will be free and some will be paid apps. "The Best New Games" section is underneath the "Best New App section".

3) Search for the app you want. For the iPad the search section is in the upper right hand corner. There is a tiny window up there with an X on the far right side of it. If you know the name of the App you want (for example Angry Birds, or 'Stache It) you can type it in here and it will take you to the app. You can also type in the kind of app you want. You could type in "Photo-Editor", "Flashlight App", or "To-Do List App". For iPods and iPhones the search button is in the lower portion of the screen.

When you find a game you like, touch it. The information for the game will pop up. This will include the game's name, the name of the developer, how many reviews and how many stars it has (the higher the better) and also the price of the game.

You will also have 3 other options to find out more about the game: Details, Reviews and Related. Below is a little bit of information about each of these:

<u>Details</u> – In Details you will see some or all of the following:

Screen Shots –There will be pictures of the app or game during game play. Use this to see if the app is appealing to you.

Description –You will find out how to use the app. If it's a game, you will learn how to play it, what (if any) levels it has, and all the reasons why the developer thinks you should play this game.

What's New – This will list all of the latest updates the developer has made to the game to make it better than the last version

Supports – This lets you know if you can post your scores to Game Center

Information – Tells you who the developer is, what category the game is in, when it was last updated, which version of the game this is, the size of the app, it's average app rating, and the compatibility (which IOS the app uses and what device the game will work on).

In App Purchases – Many free games will also have special powers ups you can purchase for a fee. You don't have to purchase them. If an app has things you can purchase within the game it will show up here. Many of the games I play contain In App Purchases however I've never bought any. You can almost always play the game without them. For more information on In App Purchases see the chapter in this book.

Version History – Lets you know of past updates

Developer Apps – This section lets the app developer show you some of the other games or apps they've made.

Developer Website – Clicking on this will take you to their website where there is usually a list of other games they've made, information on the developer and sometimes a way for your to complain or make comments directly to the developer about their games.

Privacy Policy – Lets you know how they use the information they collect from you.

Reviews and Ratings
This section lets you see all of the reviews people have left for the app. It also lets you leave a review and rating for the app.

You will see a section called "Tap to Rate" followed by five stars. If you have installed the app you and want to rate it, click on which ever star you think it deserves. If you think it deserves a four star review. Click on the 4th start. You can also click on the "Write a Review" button and leave a review of the app.

Note about reviews – When you decide to leave a review keep in mind that a real person, with real feelings made this app. And most likely they do read the reviews. Anything negative you say about them can and probably will have a negative effect on them and the app. This goes for the star rating as well.

If you're not happy with the app and it's not because something is wrong with it, but more because of personal taste, don't leave a negative review. Instead, see if you can contact the developer through email or their website (which you will find in the "Description" section of the app on the app store). As a developer, I can tell you that if there is something we can change to make the app a better experience for you, we want to know about it. We may or may not be able to make the changes but we do appreciate the feedback. Once you leave a negative review, it's on there forever.

Related

This section shows you other apps that customers have purchased. Scroll from left to right to see them.

The two most important things to watch out for are the price and the reviews. Make sure you read the reviews and pay attention to the star rating.

Buying an app - If you decide you want the app you can download it to your device. To do this, click on "Free" (if it's a free app) or click on the price of the app (if it's a paid app). When you do, the button you clicked on will change and will now say "Install App". If you want the app, click that button. A window will pop up asking you to type in your iTunes password. When you set up your device you should have set up an account and created an Apple ID and Password. If you haven't done that yet, you will need to do that before you can continue. If you've already done it, enter your password and click "OK". The app will now start installing

Note – You will need an Internet connection to download apps. If you have a spotty Internet connection you may have problems installing things.

Once the app finishes installing the button will change to "Open". You can either click on it to open and begin playing around with your new app, or you can search for another app if there are several you want to install. An icon for the app you just downloaded will show up automatically on your home screen for your device so it will be there whenever you're ready to use it.

Chapter 7 – How Do I Use an iTunes Gift Card?

When you set up your iTunes account you will be asked to put in your credit card information. This makes it easy to buy apps. When you buy an app in the app store you just click on the "Install App" button, enter your iTunes password and the credit card will automatically be charged. Or you can purchase an iTunes gift card. iTunes gift cards work in both iTunes and in the Apple App Store.

Why use an iTunes gift card?
1) Some people don't feel comfortable adding their credit card information to their iTunes account.

2) It's an easy way to keep tabs on how much you're spending in the Apple App Store or in iTunes. With an iTunes gift card you are setting a limit for yourself on how much you're going to spend. When you're using a credit card it's easy to spend a little more than you intended.

3) It's great for kids because they don't have access to your credit card information and can't rack up a lot of purchases.

You can purchase an iTunes gift card at Wal-Mart, Target, Kohl's and many grocery stores. They come in many different dollar amounts to suit your needs. When you purchase the card the cashier will activate it just like she would any gift card.

Note – If you've never set up an iTunes account you will have to do this first. This is done on your computer and is very simple. Go to iTunes.apple.com or just Google iTunes. Once there, look for the blue button in the upper right hand corner of the screen that says "Download iTunes". When you click on this it walks you through, step by step, how to open an account.

Once you have your iTunes gift card carefully scratch off the silver coating on the back of the card (some cards let you peal off the silver layer) to reveal your code.

You can either redeem your Gift Card on your computer in your iTunes account or you can do it on your iPhone, iPad, or iPod.

To redeem a gift card on your iPhone, iPad or iPod:
Click on the Apple App Store icon from your home screen. It's a blue icon with a large white "A" in the middle of it. When the app store comes up scroll all the way to the bottom of the page.

At the very bottom you will see a button that says "Redeem". Clicking on it brings up a window that tells you to enter your gift card or your download code. Type in the code on the back of your gift card and hit the "Redeem" button in the upper right hand corner of the little window. Another window will pop up asking you to put in your iTunes password.

Note – Even though the code on the back of your gift card is in all caps, you don't have to type it in that way. Lower case letters work.

To redeem a gift card on your computer:
Click on either the Apple App Store icon (blue with a large white letter "A" in the middle) or the iTunes icon (Blue with

a large white music note in the middle). You can type in your code from your gift card on either one of these and you will be able to buy music from the iTunes store and apps from the Apple App Store.

Once you're in either the app store or the iTunes store, look to the right of your screen for a section called "Quick Links". Under this section you will click on the word "Redeem". A window will pop up asking you to type in your iTunes password.

After you type in your password another window will pop up. You have two options. You can either type in your code that is on the back of your card or, on the newer gift cards, you can take a picture of it. If you choose the "Use Camera" option (and have a camera on your computer or laptop) your camera will turn on. You will have to position your card so it shows up in the little window. It will automatically enter your code for you.

Once your code is entered you can now shop in either iTunes or the App Store!

Chapter 8 – How Do I Search For and Buy Apps on Google Play for Android Devices?

Before you get started you need to set up a Google Play account. Go to https://play.google.com/store and click on "Sign In". From there you can create a new account by following the directions they provide.

Searching for apps in the Google Play store is pretty simple.

When you click on the "Google Play" icon (it looks like a white shopping bag with a colorful triangle in the middle of it) you will be taken to the app store's home page.

From here, there are a few ways to search for apps:

1) Check out the different categories. Google lists a few for you "Apps", "Games", "Movies and TV", "Music", "Books", and "Newsstand".

Note – When you click on one of these categories you're taken to a new screen where you have even more options to refine your search. At the top of the screen you will see the following menu options "Categories", "Home", "Top Paid", "Top Free", "Top Grossing", "Top new paid", "Top New Free" and "Trending". This menu scrolls from side to side you will need to scroll from left to right with your finger to see all of these choices. As you're swiping from left to right, a list of apps shows up underneath each category. For example if the"Top new Paid" category is highlighted, there will be a list of apps under it to choose from that are the most popular new apps in the Google Play store at the moment.

2) Scroll up and down to see which apps Google recommends. Some of these will be the most popular apps at the time and some will be ones that Google thinks you'd like.

3) If you know what app you want, you can search for it by clicking on the picture of the magnifying class at the top of the screen. Once you click on it you can type in the specific app you're look for (Angry Birds) or you can type in the kind of app you're looking for (Flash Light App or To Do List App).

When you find an app you think you like, click on it. You will be taken to the "Details" page. On this page you will see some or all of the following depending on the app:

Screen Shots –There will be pictures of the app or game during game play. Use this to see if the app is appealing to you.

Description –You will find out how to use the app. If it's a game, you will learn how to play it, what (if any) levels it has, and all the reasons why the developer thinks you should play this game.

<u>What's New</u> – This will list all of the latest updates the developer has made to the game to make it better than the last version

<u>Reviews and Ratings</u>
This section lets you see all of the reviews people have left for the app. It also lets you leave a review and rating for the app.

Note about reviews – When you decide to leave a review keep in mind that a real person, with real feelings made this app. And most likely they do read the reviews. Anything negative you say about them can and probably will have a negative effect on them and the app. This goes for the star rating as well.

If you're not happy with the app and it's not because something is wrong with it, but more because of personal taste, don't leave a negative review. Instead, see if you can contact the developer through email or their website (which you will find in the "Description" section of the app on the app store). As a developer, I can tell you that if there is something we can change to make the app a better experience for you, we want to know about it. We may or may not be able to make the changes but we do appreciate the feedback. Once you leave a negative review, it's on there forever.

<u>Users Also Installed</u> – This section shows you the apps that users also installed.

Similar Apps – These are apps that Google things are similar to the one that you are currently looking at.

Developer – This tells you who the developer is and may also include their email address and website. It may also tell you what category the game is in, when it was last updated, which version of the game this is, the size of the app, it's average app rating, and the compatibility (which OS the app uses and what device the game will work on).

Google Play Content – This will only appear on game apps. This lets you know if you can post your score to Google Play and see the scores of other players.

In App Purchases – Many free games will also have special powers ups you can purchase for a fee. You don't have to purchase them. If an app has things you can purchase within the game it will show up here. Many of the games I play contain In App Purchases however I've never bought any. You can almost always play the game without them. For more information on In App Purchases see the chapter in this book.

Version History – Lets you know of past updates

Developer Apps – This section lets the app developer show you some of the other games or apps they've made.

Developer Website – Clicking on this will take you to their website where there is usually a list of other games they've made, information on the developer and sometimes a way for your to complain or make comments directly to the developer about their games.

Privacy Policy – Lets you know how they use the information they collect from you.

The two most important things to watch out for are the price and the reviews. Make sure you read the reviews and pay attention to the star rating.

Once you decide to purchase the app, click on the purchase price. If the app is free click on the word "Free" or on some apps "Install". A window will pop up with the permissions list. To learn more about permissions see the chapter on permissions in this book. Most of them aren't as scary as they may seem. If you still want to download the app click "accept". Another window will pop up showing you the amount you have in your account and the price of the app. Click on "Buy". You will have to enter your password. If you haven't set up an account yet, you can do it online at the Google Play website.

Note – To pay for apps you can either buy a Google Play Store Gift Card (see next chapter on how to redeem) or you can purchase Google Play Credits from the Google Play website after you've logged into your account.

Chapter 9 - How Do I Use a Google Play Gift Card?

You can purchase a Google Play Gift Card at places like Wal-Mart, Target, and even grocery stores. They come in a wide range of amounts to suit your needs.

Why use a Google Play Gift Card?
1) Some people don't feel comfortable adding their credit card information to their Google account.

2) It's an easy way to keep tabs on how much you're spending in the Google Play Store. With a gift card you are setting a limit for yourself on how much you're going to spend. When you're using a credit card it's easy to spend more than you intended.

3) It's great for kids because they don't have access to your credit card information and can't rack up a lot of purchases.

Redeeming a card is simple. You can either redeem your card online by going to the Google Play website or you can redeem it on your smartphone or tablet.

To redeem your gift card on your computer:
You will need to set up a Google Play account. Go to https://play.google.com/store. In the upper right hand corner you will see "Sign In". When you click on that you will have the option to either sign in or create a new account. Follow the directions on the website to set up an account.

Once you have your account set up go back to the home page for the Google Play Store and look on the left hand side of the screen. In the menu look for the word "Redeem". When you click on it a small window opens and asks you to enter your gift card or promo code. On the back of your Google Play Gift Card carefully scratch off the silver coating revealing your code. Enter the code in the space provided on your computer then click on "Redeem".

To redeem your gift card on your device:
To redeem a gift card on your phone or tablet find the Google Play icon on your device. When you click on it you're taken to the Google Play Store. From here you will need to look at the top of your screen. Look for three dots or three small lines all lined up on top of each other. That's called the action overflow menu. On some devices it's on the upper right hand side and on some it's on the upper left. When you find it, click on it and a menu will pop up. Click on "Redeem". Another window pops up and asks you to enter your code. Scratch off the silver coating on the back of your Gift Card revealing your code. Type it into your smartphone or tablet. A new window will pop up letting you know you've successfully added money to your Google Play account.

Chapter 10 – How Do I Search For and Buy Apps on Kindle?

Apps are easy to purchase on your Kindle Fire. From your home screen you'll find a list of things to click on at the top of your screen. Look for the **"Apps"** button and click on it. This takes you to a screen with all of the apps you've installed (or that came installed) on your device.

In the upper right corner you will see the word "**Stor**e". Clicking on it will take you to the app store. You have a couple of options:

1) Search the Appstore – At the very top of your screen in the middle you will see a little window that says "Search Appstore". If you know the name of the app you want you can type it in here (click on the words "Search Appstore first). You can also use this option if you know the type of app you want. For example if you want a free calculator app simply type that in. Or you could type in "task manager" to get a list of apps that act as to-do lists.

2) Look through the categories Kindle offers – Below the section to search for apps, but still towards the top of your screen, you will see a list of categories. Here are the ones currently on Kindle Fire (keep in mind that as Kindle makes updates this could change): Best Sellers, Games, New Releases, Test Drive and All Categories.

The Test Drive feature is a neat way to try out an app before you buy it. If you click on "Test Drive" a list of apps to choose from will appear. Scroll through the apps by scrolling up or down. When you find one you want to try out, click on it. A description of the game pops up along with screen shots of the game. If you want to test-drive it click on the green "test drive" button by the game icon. You will be allowed to play the game or try out the app for 10 minutes. If you decide you don't like the game and don't want to test-drive it anymore you can click on "quit" in the upper left hand corner. If you like the app and want to download it click on the price or the "Free" button in the upper right hand corner.

Note – I'm not sure why but some of the apps in the "Test Drive" section don't actually have a way to test-drive it. Also some of the apps I tried to test drive crashed and didn't work. If the app is free you might as well install it and then get rid of it if you don't like it.   I was able to try out a few of the games I clicked on and I think it's a neat feature when it actually works.

3) Look through the "Featured Apps and Games" – These are apps that Kindle has decided to promote. You can scroll by sliding your finger to the left (or to the right if you want to see the ones you've already passed).

Note – If you scroll down further you will see other apps that Kindle is recommending for you.

When you find an app you think you like, click on it. You will be taken to the "Details" page. On this page you will see some or all of the following depending on the app:

Screen Shots –There will be pictures of the app or game during game play. Use this to see if the app is appealing to you. Scroll from left to right to view all of the screen shots.

Description –You will find out how to use the app. If it's a game, you will learn how to play it, what (if any) levels it has, and all the reasons why the developer thinks you should play this game.

Key Details – This section is to the right of the description. It lets you know if there are in app purchases and also if you can post your score to Game Circle. It will also tell you if the app is for all ages or if it's appropriate for older users (in this case it might say "Guidance Suggested")

Customers Who Bought This Item Also Bought – This is a list of other games customers have purchased. Scroll from left to right to view all of them.

Reviews and Ratings
This section lets you see all of the reviews people have left for the app. It also lets you leave a review and rating for the app if you have purchased it.

Note about reviews – When you decide to leave a review keep in mind that a real person, with real feelings made this app. And most likely they do read the reviews. Anything negative you say about them can and probably will have a negative effect on them and the app. This goes for the star rating as well.

If you're not happy with the app and it's not because something is wrong with it, but more because of personal taste, don't leave a negative review. Instead, see if you can contact the developer through email or their website (which you will find in the "Description" section of the app on the app store). As a developer, I can tell you that if there is something we can change to make the app a better experience for you, we want to know about it. We may or may not be able to make the changes but we do appreciate the feedback. Once you leave a negative review, it's on there forever.

Permissions – This tells you what you are giving the app permission to do or what information you are letting them gain access to. To learn more about permissions see the chapter in this book.

Developer – This tells you who the developer is and may also include their email address and website.

Product Details – This tells you the rating of the app (if it's for all ages or if guidance is suggested), the language the app uses, the size of the app, the app's ASIN (this stands for "Amazon Standard Identification Number), the original release date, and the current version of the app. Sometimes it will also include the developer's privacy policy.

Purchase Details – Here is where you can find Amazon's return policy

More Apps By This Developer – This section lets the app developer show you some of the other games or apps they've made.

Developer Website – Clicking on this will take you to their website where there is usually a list of other games they've made, information on the developer and sometimes a way for your to complain or make comments directly to the developer about their games.

Privacy Policy – Lets you know how they use the information they collect from you.

The two most important things to watch out for are the price and the reviews. Make sure you read the reviews and pay attention to the star rating.

How to purchase an app:
When you are in the details screen of the app, click on the price of the app, which should be located on the left hand side of your screen under, or next to, the app's name.

If it's a free app, when you click on it another little button should appear that says "Get App". When you click on this your game will download and an app icon will automatically be added to your home screen.

If the app is a paid app, click on the price of the app to make your purchase. To find out what Amazon coins are, or to learn how to use an Amazon or Kindle gift card, see the following chapters.

Chapter 11 - What are Amazon Coins For Kindle?

Sometimes when you purchase an app you will receive Amazon Coins. To see if an app you want comes with coins look around the price of the app. It's usually located just above the price. It will say something like "Earn 60 coins when you buy this app".

When you install the app the coins will automatically be added to your Amazon Coin Account. You can use the coins to purchase other apps. To see how many coins you have look in the lower right hand corner of your screen when you're in the Appstore. If there isn't an amount there you can check your account on Amazon.com

Each coin acts like a penny so if a game costs $1.99 you use 199 coins. Not all apps will let you pay in this manner but many do. If the app allows Amazon Coins as payment it will say so right next to the price. For example it will say ".99 or 99 coins"

Don't have very many coins? No worries, you can buy more! If you're in the app store and you see the amount of coins you have in the lower right hand corner, click on it. A window will pop up and let you purchase Amazon coins. Doing this will actually save you a little bit of money. 500 Amazon Coins will cost you $4.80. This saves you 4%. The more you buy, the more you save. For example if you buy 2,500 coins it will cost you $23.00, which is an 8% savings. You can also buy them from your account on Amazon.com

Chapter 12 - How Do I Use an Amazon or Kindle Gift Card?

You can use either an Amazon gift card or a Kindle gift card to purchase apps and books for your Kindle. You can buy them at Wal-Mart, Target, grocery stores and other department stores.

Gift cards are simple to use. You can redeem them on your Kindle or online at Amazon.com. In this chapter we're going to focus on how to redeem your card using your Kindle.

1) Make sure you're in the Kindle Appstore (click on "Apps" on your home page then click on "Store" in the upper right hand corner of the screen that pops up).
2) Look at the bottom of your screen. In the middle there should be a little menu button that looks like half of a rectangle with three lines in it. Click on it.
3) When the menu pops up, look for the selection that says "Gift Card and Promos". Click on it.
4) Another screen will come up that says "Redeem a Amazon.com Gift Card or Promotional Code." Flip your card over and carefully scratch off the silver area to get your code. Type your code in and hit "Redeem". A little notification should pop up and say that you were successful in redeeming your card.

Sometimes when I type in my code the money doesn't appear in my account right away. It may take a few seconds. You may have to go out of that screen and then go back to it to see that the amount has been added.

Note – When typing in your code on the back of your card you don't have to type in the dashes or use capital letters. What you do want to watch out for is auto correct. I had a terrible time typing in my code because Kindle wanted to make sense of what I was typing. I had the letters "A", "P", and "S" next to each other in my code and Kindle wanted to change it to "Apes". You may want to turn off auto correct before you type in your code.

How to turn off auto-correct on your Kindle Fire – First, swipe down from the top of your screen to access the notification bar. Click on the "**More**" button. Look down until you see "**Language and Keyboard**" and click on it. A screen pops up with a few options. Click on "**Keyboard**". Next click on "**Keyboard Settings**". On the next screen you will see the option to turn of "**Auto-Correct**". If you decide later you want it back on, just go through these steps again and switch the controls to "On".

Now that the money is in your account you can purchase an app! When you find an app that you want to purchase, click the button that shows the price. A window will pop up with an option to purchase the app using your Amazon coins or using actual money. If you have enough coins in your account you can use those or use the money from the gift card you just redeemed. After you click on your payment method click the green "**Get App**" button. The app will immediately start to download. You can see its progress by watching a blue bar extend across your screen. Once it's installed you can click on the orange "**Open**" button to use your app. If you're not ready to use your app quite yet that's OK. Whenever you're ready to use it you can click on the app icon that is automatically added on your home page.

Note – To check the balance on your account click on the "Gift Card and Promos" button from the menu just like we did to redeem your card. Towards the top of screen it will say "Amazon.com Gift Card account balance" followed by the amount you still have available.

Chapter 13 – What if I Bought an App on Accident?

For Apple Apps– Apple makes you put in your password each time before you buy an app so hopefully that will stop you from buying an app accidentally. Apple says that all app purchases are final but there is a chance you can still get your money back.

You will need to go to your iTunes account on your computer (not your phone or tablet). Log in with your apple ID and go to your purchase history. There should be a "report a problem" link. If you click on that and tell them that you bought the app by accident (maybe you purchased the wrong thing) they may give you your money back. It could take a few weeks for them to refund your money.

For Android Apps – Android makes it easy to get your money back if you bought an app on accident. They give you 15 minutes after you purchase the app to get a refund.

If you've decided you don't want the app, or you bought it by accident, click on the "Refund" button. If the button isn't there then your 15 minutes has passed. Once you click on the button just follow the directions and your money will be refunded and the app will be removed.

If you waited longer than 15 minutes Android says you will have to contact the developer and ask for a refund. It's up to them to decide to return your money or not.

For Kindle Apps – Just like with Android there is a small window of time you can ask for a refund. If you accidentally buy an app on Kindle go to the Kindle Customer Service and click on "**Help**". After that, click on "**Kindle Help**" and then "**Contact Us**". They may change the menu options on the site but the key thing to look for is "Contact Us". Once there you can ask them to "**Call You**". As surprising as it seems (at least to me) they call you right away. You can then explain your situation and ask for a refund.

Chapter 14– What are "In App Purchases"?

In app purchases are power ups or add-ons to games. Most of the time they are offered on free games but occasionally you will see them on paid games as well.

You will be able to tell if a game contains in app purchases by looking at the "Description" or "Details" section of the app before you buy it.

An example of a game that has in app purchases is the Stache' It app. It's a fun app that lets you download a photo of yourself or a friend and then put mustaches and beards on them. You have several free items to choose from. If you get bored with those, or just want more options, you can purchase other items to add to your pictures. For .99 you can buy a package that has several lips to add to your picture or you can purchase a package that has several crazy eyes.

Some games have in app purchases that let you buy more lives if you die in a game, give you extra levels, or let you purchase items for your game.

An example of a game that lets you purchase extra lives is the very popular game Candy Crush. It's similar to Bejeweled but instead of gemstones, the board is full of candy. Each level has a different goal. Sometimes it's to clear the board in a certain amount of time, other times it's to finish it in a certain number of moves. You are only allowed to fail five times. After that you either have to pay to add more lives (that's the in app purchase) or you have to wait 20 minutes for another life to be added for free. The game can be played without using the add-ons. However, some people really enjoy the game and don't want to wait for the free lives to appear again in 20 minutes so they purchase the add-on to get more lives right away.

Most games that have in app purchase can still be played without purchasing them.

Chapter 15- What's up with all of Those Ads?

The developer of the app spent many hours (hundreds on some of the larger apps) and lots of money making the app and of course they would like to make their money back (and then some).

There are a few ways the developer gets paid:

Charge for the app – Some apps on the app stores are .99 while others are pretty pricey. They usually don't contain any of those annoying ads that pop up when you open the game or when you finish a level.

Put ads on the game – Ads are put on the game in the form of banner ads (a banner at the top or bottom of the game), a full-page ad, or even video ads. These ads show you other games or apps you can download. If you were to click on one of these ads the developer makes a little bit of money (sometimes just a few cents but it all adds up – no pun intended). Sometimes developers go a little overboard with the ads and they pop up every few seconds. This can get very annoying. Other developers try to add them in at less irritating times (not during game play).

The ads are the price you pay for a free game. Many times there is also a paid version of the game that doesn't contain ads. Usually what I do is purchase the free game to see if I like it. If the ads don't bother me I'll keep playing the free version. If I really like the game and don't want to see the ads, I'll get the paid version.

Include In App Purchases - The developer also gets paid by using the In App Purchases we talked about in the last chapter.

One in app purchase to watch for is the "Remove Ads" option. Usually for .99 you can upgrade your game (or other app) so it doesn't include ads.

Chapter 16 –Is There a Difference Between the Free Version of an App and the Paid Version?

Free games almost always have ads and/or in app purchases. They can also be free trials or shorter versions of a paid app. To get this information look in the "Description" or "Details" section of the app in the App Store.

The paid versions usually don't contain ads and sometimes they might contain more information or more levels (if it's a game) than the free version.

Chapter 17 – What Happens if I Accidentally Click on an ad?

The ads are sometimes strategically placed so you accidentally hit them while you're using the app. You will accidentally hit one at some point. It's ok if you do. You don't automatically buy the game or app in the ad. Instead it takes you to the app store where you can buy the game if you want to.

To get back to your game you will have to go back to your home screen on your device and click on the app icon again.

Chapter 18 - What's the Safest Way to Buy and Bank on my Phone or Tablet?

If you're out shopping and you want to know how much is left in your checking account it might be tempting to pull up the internet and access your banking information on your phone or tablet the same way you would at home on your computer. However, that's not the safest way to check your balance.

The safest way to bank on your phone or tablet is to download your bank's mobile phone app. Ask your bank if they have one or go to your bank's website and search for a link there for their app. Once you click on the link it should take you to the App Store where you can download the app.

It's better to go to your bank's website and look for the link to their app instead of just searching for it in the app store. At one point there were over 50 fake bank apps in the Android app store. Apple carefully looks through each app that is submitted while Android doesn't. Android apps are approved automatically without any scrutiny. You are more likely to find a fake app on Android than you are on the Apple store.

If you don't want to use your banks app and insist on using your Internet browser make sure you don't use public Wi-Fi. Use your phone provider's network. It's much harder for hackers to tap into it.

The same guidelines work for shopping on your phone or tablet. Don't make purchase while using Wi-Fi. Also, look for apps from the store. EBay and Amazon both have apps you can download that make it more secure for shopping on your phone or tablet.

A note to Android users – Before you download a banking app or one from a store, make sure to read the reviews. Like I mentioned before Android doesn't scrutinize the apps that developers submit so you have a greater chance of being scammed. Reading the reviews of apps will bring your attention to any problems or warnings other users are reporting. The safest way to get your banking app is to click on the link from your bank. The same goes for stores.

Chapter 19 - Can I Remove Apps From my Device?

At some point you may decide you don't want a certain app on your phone or tablet anymore. Or maybe you will need to get rid of it because you're running out of storage and want room to download other apps. No problem! Removing apps from your device is simple once you know how to do it.

Removing apps from Apple Devices – Touch the app icon you want to remove and keep your finger on it. If the app is removable (some apps that come already installed on your device aren't removable) a little X will appear at the corner of the app. Click on the X and a window will appear asking if you're sure you want to delete that app. You can either click "cancel" if you don't want to delete it or click "Delete".

Note – To get out of "Delete Mode" click on your home button.

If you decide later you want the app back just log in to your iTunes account and look under your Purchase History. You will be able to find your app and put it back on your device.

Removing apps from Android Devices – The way you delete apps depends on what version of the operating system you have.

For versions before Ice Cream Sandwich:
1. Tap on the Menu button (either a hard or soft button)
2. Tap on Settings: Applications: Manage applications
3. Tap on the app you want to delete
4. Tap on Uninstall

For versions after Ice Cream Sandwich including Jelly Bean:
1. Find the app icon you want to delete and touch it.
2. Keep your finger on the app until the screen changes. You will see a menu at the top of your screen.
3. One of the options at the top of the screen in the upper right hand corner is garbage can. Keep your finger on the icon and drag it up to the garbage can.
4. Let go of your finger and the app will be deleted.

How do you know which Android OS (operating system) your device has? Go to "Settings" on your phone and click on either "**About Phone**" or "**About Device**" (your device will say one of those two). Find where it says "**Android Version**". Look at the number displayed underneath it.
4.0= Ice Cream Sandwich
4.1, 4.2 and 4.3 = Jelly Bean
4.4 = Kit Kat

Removing apps on your Kindle:
1. Touch the icon of the app you want to delete. Hold your finger on the icon.
2. A window pops up and asks you if you want to delete the app from your Kindle.

Chapter 20 - How Can I Make Sure my Kids or Grandkids Don't Purchase Apps on my Smartphone or Tablet?

There are a couple of ways to do this. No matter what kind of device you have (Kindle, Apple, or Android product) you can turn off your Wi-Fi. With out Wi-Fi they can't download any game from any app store, not only that, they can't even get into the app store. You can turn off your Wi-Fi in your settings. A word of caution – kids these days are really smart and super sneaky. They can figure out how to turn Wi-Fi back on quicker than it took you to figure out how to turn it off. While adults have to take time to learn all of this new technology I think kids today are naturally born with this knowledge. Turning off Wi-Fi will probably only work with younger kids.

Another way to protect your account is to never give your child or grandchild your password. In Android and Kindle there is a way to disable your password so you never have to type it in when you want to buy something. Make sure you have your password enabled on your account so others aren't spending your money. If someone wants to purchase a game have them hand you the device and you enter your password so they never know what it is.

How to make sure a password is needed to make a purchase:

For Android devices - Go to the Google Play Appstore by clicking on the icon on your home page. Once there hit your menu button on your phone or tablet. A little menu button will pop up and you to tap "**Settings**". When you do this, a new window will open and you will want to scroll down and until you see "**Password**". If you want to make sure that it asks for your password each time someone tries to make a purchase, make sure the box is checked. If, for some reason, you don't want to have to put in your password then uncheck the box. If you do this a box will pop up asking you to verify that you don't want to enter a password anymore and letting you know that you assume all responsibility for anything that is purchased.

For Kindle Fire - Swipe down from the top of your screen to access the notification bar. Click on "**More**". A "**Settings**" menu will pop up. Look for the words "**Parental Controls**". When you click on it you will be brought to a screen that lets you do a couple of cool things.

First of all you can click on "Parental Controls" and you will be asked to set up a password of at least 4 characters. Once you have your password set up you can block specific things you don't want your kids see. You can also make it so you have to have a password to turn on Wi-Fi, purchase apps, or watch videos.

The other thing you can do from the main "Parental Controls" page is set up a profile for your kids. Click on the "Kindle Free time" button and you will be walked through how to set up a profile for your child. Once the profile is set up you can add books, TV shows, movies and Apps for your kids. You can also set a timer to restrict the amount of time they play on the Kindle each day.

For Apple devices - When you first make a purchase it asks for your password. After that you can buy whatever you want within 15 minutes and not have to reenter your password. Apple did this to make things easier for you when you have several songs you want to download. If you're downloading a list of songs it can get irritating having to enter your password each time.

Little kids accidentally learned how to take advantage of this. The parent or grandparent purchases a game for them and hands the device over to the child. The child plays the game but quickly runs out of coins. A window pops up and says "Buy more coins for this game! Click here!" The kid clicks the button and, since it's been less than 15 minutes since the game was purchased, the transaction goes through without asking for a password.

Apple has learned of this problem thanks to thousands of angry parents across the nation and a hefty lawsuit they brought against Apple. Thankfully there's a way you can get rid of the 15-minute window.

Here's how to do it:
Go to "**Settings**". The icon will be on your home page and will have a picture of gears in it. A new window will pop up and you will look on the left hand side for the word "**General**". Click on it. On the right hand side of the screen you will see several options. You're looking for the word "**Restrictions**". To the left of the word restrictions you will see either the word "**On" or "Off**". If it says "Off" (which it

probably will if you've never done this before) a new window will pop up but it will look a little hazy. You won't be able to click on anything other than the words "**Enable Restrictions**" at the top of your screen.

When you click on "**Enable Restrictions**" a new window will pop up that says "**Set Passcode**".   You will have to type in a 4-digit number for your code. Make sure to pick a code you will remember but others in your family may not think of. You don't want to chose the month and day of your birthday and then have your kids or grandkids figure it out. Once you type in the 4 numbers you've chosen it will ask you to re-enter it. Now the haziness of the restrictions screen is nice and clear and you can click on things. Look down your list of options until you find "**Require Password**". Out to the side it probably say "**15 minutes**". If you click on this you will be taken to another window where you will have two options: "**Immediately**" or "**15 minutes**". If you want it to require a password every time you purchase something click on "**Immediately**". There isn't a save button (it will automatically save) so hit the blue "Restrictions" button at the top left of your screen to take you back to the "Restrictions" page.

Another handy thing that Apple does is it takes away the ability to purchase items from the iTunes and App store. If you go into "**Restrictions**" just like we did above look for the list of items under "**Allow**". There are about a dozen items listed and each should have a little green switch next to it. If you don't want anyone to purchase anything at all you can click on the switch next to "Installing Apps" and "iTunes Store". When you do this the icon on your home page will disappear. When you want it back just go back to

"**Settings**", "**General**", "**Restrictions**" (you will have to enter your 4 digit code at this point) and then click on the switch turning it back to green and putting the icon back on your home page and giving you the ability to make purchases again.

Note – You can also switch off the ability to delete apps so you don't have to worry about kids deleting things you want. You also have the ability to turn off in-app purchases.

It's ok to play around with this section to see how all of these things work. I know when people are new to apps or smartphones or tablets they can worry about screwing something up. As long as you remember your passcode you will always be able to go in and undo whatever changes you've made.

Chapter 21 – What do all of Those Words Mean?

<u>3G</u> – This means "3rd Generation" and Verizon.com explains it this way:

"3G is shorthand for "3rd generation," and refers to a networking standard in cell phone technology that is capable of providing high-speed data service to mobile devices."

If your phone has 3G it means it can access the Internet (either through Wi-Fi or through your phone's provider). It's based on speed (how quickly you can access and download information from the internet).

<u>4G</u> – This means "4th Generation" and it's faster than 3G. It's not quite as fast as if you were home, working on your computer using the Internet, but it's still 10 times faster than 3G. Not all phones come configured with 4G so if speed is important to you, make sure you look for this when you upgrade to a new phone or tablet.

<u>4G LTE</u> – The LTE part of this stands for "Long Term Evolution". It delivers the fastest and most reliable Internet connection possible (as of right now). Some people refer to is as "true 4G". Using 4G LTE is just like using your Internet at home.

<u>Wi-Fi</u> – What do those letters stand for? Some believe it's "Wireless Fidelity" but they would be incorrect. It turns out the term "Wi-Fi" was just a catchy name that doesn't actually stand for anything. Don't believe me (I don't blame you. It seems ridiculous to me too)? Check out this site:

http://www.scientificamerican.com/article/pogue-what-wifi-stands-for-other-wireless-questions-answered/

So what's the definition? It allows your computer, smartphone or tablet to connect wirelessly to the Internet.

Disabled – This means you've turned something off or gotten rid of the ability to do something. For example if you disabled "Password Protection" on your device it means the device won't ask you for a password to do things.

Enabled – This is the opposite of "Disabled". You are turning something on or activating a feature on your device.

Smartphone – A cell phone that can do many of the same things your computer can do – access the internet, take pictures, show videos, etc.

Tablet – They are like small computers. They don't have near the memory or speed that computers have but they can function in much the same way. You can even purchase keyboards to make them function even more like a computer.

IOS – This is the Apple operating system. The "I" in the front is just like the "I" in iPod, iPad and iPhone. The "O" and the "S" mean "Operating System". What does the "I" stand for? Some people think it stands for "internet". Apple is currently at IOS 7. Each time they update their operating system your phone will have more features or it will make the current features easier to use. That's their plan any way. For those that don't like change, an operating system update can be very frustrating until you get used to it.

SMS – Short Message Service (texting)

MMS – Multi Message Service (sending videos, pictures or sounds)

SD Card – They are tiny little memory cards in your phone or tablet. You can save pictures, videos, documents and your contact list to them. When you get a new phone your contents can be transferred from your old SD card to the new one in your new phone. That way you don't lose any of your photos or information.

Chapter 22 – Can I Have Some Final Words of Wisdom?

Hopefully you now have a better understanding of how to use apps. The more you use them, the more comfortable you will feel with them. There are a lot of fun and helpful apps out there – some can make our lives easier or more organized and some can keep our minds busy while we wait in the waiting room for our dentist appointment. Some apps keep us closer to our far away friends and family and others keep us informed about the weather.

So what should you do now? How about the following:

1) Sign up for my monthly newsletter - If you'd like to keep up to date on app news, you can check out my website that I made just for you! Every month I will post a new article about what's new in the app world. You can use this knowledge to have conversations with your kids, grandkids and co-workers so you can sound just as smart as they do! I'll also post an article about an app or two each month that is popular at the time and let you know what the permissions mean for each app as well as a description. Once a month I'll send you a quick email with a link to the articles I post, along with a description of what I posted that month.

If you'd like to sign up for my monthly newsletter you can go to my website **WhatAboutApps.com** and sign up on my homepage.

2) If this book was helpful leave me a good review on Amazon.com. I am a self-published author and those good reviews mean a lot to me! It's always nice to know that I helped someone.

3) If there is a question you think should be in this book just ask me. You can go to my website WhatAboutApps.com and find the "Contact Me" section.

4) Don't be afraid! A big thing to remember at the beginning of your app journey is that it's unlikely that you are going to mess something up. Just use your newly acquired app knowledge and you will be just fine.

Chapter 23 – Can You Give me a Few Blank Pages so I Can Take Notes?

Sure! But only because you asked so nicely!

Notes:

Notes:

Notes:

Notes: